HOW DO YOU SURVIVE ON AN ICEBERG?

AND OTHER PUZZLES WITH SCIENCE

THIS IS A CARLTON BOOK

First published in 2015 by Carlton Books Limited
an imprint of the Carlton Publishing Group
20 Mortimer Street
London W1T 3JW

10 9 8 7 6 5 4 3 2 1

A catalogue record for this book is available from the British Library

ISBN 978 1 78097 670 9

Printed in Dubai

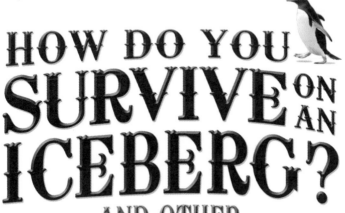

HOW DO YOU SURVIVE ON AN ICEBERG?

AND OTHER PUZZLES WITH SCIENCE

ERWIN BRECHER

CARLTON
BOOKS

Contents

About the Author   6

Introduction   7

Puzzles   9

Answers   99

About the Author

Erwin Brecher was born in Budapest and studied mathematics, physics, psychology and engineering in Vienna, Czechoslovakia and London. His first book was published in 1994. Erwin wrote many more books on puzzles, IQ, chess, bridge and scientific subjects. In September 1995 he was awarded the Order of Merit in Gold from the city of Vienna in recognition of his literary achievements.

Introduction

Hello and welcome to *How Do You Survive On An Iceberg?*. The short answer to the cunning conundrum posed in the title of this puzzle-setting tome is, perhaps, "not terribly well" or even "miserably". And they're both right answers, for sure. But the long answer, the really spot-on solution that we are searching for is much more fascinating and complex, one filled with tricky concepts, mind-bending ideas and brain-busting perceptions that question the very fabric of existence. And, like every riddle in this book, the harder and more complex the challenge, the increased excitement you'll feel when it is ultimately cracked open and solved.

From enigmatic paradoxes to eyeball-tickling geometry, energy equations to the unbelievable choreography of fish in the dancing waters of any ocean, this book will have you scratching your head quicker than you can say "sphygmomanometer" (see page 14). But don't take our word for it. Turn to page 9 now and begin the adventure of decoding and demystifying some of the most entertaining puzzles ever devised by humankind. OK, they're not that difficult... but they're not for the lazy-minded, either – you have been warned! Put on your thinking cap (if you've got one), find a comfy chair (some of the challenges may knock you off your feet!) and power on through until you reached the end. There's no time limit to any of the riddles, but the quicker you work them out, the greater sense of acheivement you'll feel.

A word of warning: *No peeking at the answers* from page 99, until after you think you've solved the puzzles. Then, and only then, are you allowed to check.

There are lots of mysteries and riddles in store for you, so let's not delay any longer. So, tell us, *how* do you survive on an iceberg...?

HOW DO YOU SURVIVE ON AN ICEBERG?

There are many stories of shipwrecked sailors dying of thirst or being driven insane by drinking seawater. Eskimos have no source of freshwater. Does polar ice contain salt, and if so how are the Eskimos dealing with the problem? Would you be able to survive on an iceberg?

Solution on page 100

HEAD
SCRATCHER

Project yourself back more than 2,000 years in history. You are a scholar at the court of Phillip V, King of Macedonia. It is already known that the Earth is a sphere, and you are commissioned by the king to measure the circumference of our planet.

How would you go about it?

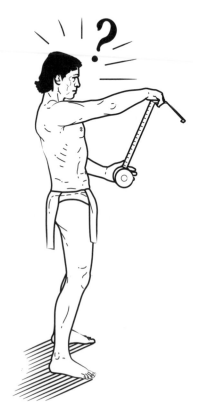

Solution on page 100

A SCHOOL
OF FISH

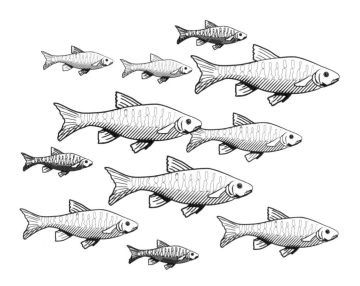

Some of my favourite TV programmes are wildlife or nature programmes.

Have you ever wondered how schools of fish, or flocks of birds, move in unison almost as if choreographed?

Is there any instant communication between the lead animal and followers indicating a change of direction, or is there a more plausible explanation?

Solution on page 101

TREE PARADOX

Trees are one of nature's most beautiful creations. They are as varied as the fauna in size, appearance and habitat. Some are giants. The Mexican Montezuma cypress has an average diameter of about 12 metres (40 feet) but can grow to 43 metres (140 feet) tall. Some redwoods are even taller!

All trees depend largely on their roots for water.

Yet we have learned at school that the maximum head of water that a suction pump can achieve is about 10 metres (33 feet).

How, then, can water reach the treetops of redwoods?

Solution on page 101

KNUCKLES

Are you one of those who, by habit, pull their fingers from time to time, to hear the knuckles crack? If you are, you've probably noticed that having heard the crack, you cannot repeat the performance – you will not hear a second crack until much later.

Can you think of an explanation?

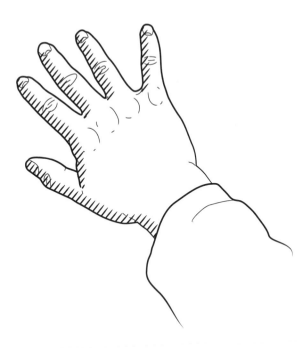

Solution on page 102

UNDER PRESSURE

Blood pressure is a measure of the pressure created in the circulatory system by the heart as it pumps blood through the main arteries.

The instrument (above) used to measure blood pressure is called a sphygmomanometer. It is always recommended that the blood pressure be taken with the arm at heart level. Why?

Could the pressure not be measured on a leg?

Solution on page 102

SEEDS OF TRUTH

When I was young my parents encouraged me to plant seeds that I found in my food. This had the effect of encouraging an interest in the natural world, gardening and food. I had great success with many things – cucumbers, melons, apples, strawberries, and even oranges on occasions. Not bad for the temperate climes of a London garden.

On one occasion I even tried a coconut that I won at a fair. This was one of my failures. But why? Is a coconut a seed? Was the temperature too cold for germination? What was the reason?

Solution on page 102

PEST CONTROL

This year, the beautiful roses I grow in my garden have greenfly. Luckily the infestation is not as bad as in previous years. I notice that there are some ladybirds and lacewings around also, and they eat the aphids, thankfully keeping the whole mini ecosystem under control. The situation seems fairly stable, but I worry about the damage they must be doing to my plants. Every summer I look forward to these roses blossoming in full bloom and turning my garden into a veritable feast of colour and life.

I could apply some pesticide – that would kill some greenfly, for sure – but it could also damage the beneficial insects, too.

My question is this: if I have a stable population of greenfly and their predators and I used a spray that kills the same proportion of both, would I then have a smaller, but still stable population?

Solution on page 103

BOXING MATCH

Take a paper match out of a book of matches and throw it into the air. It will invariably land, as one would expect, on its flat side.

Can you find a way to make the match land most of the time on its edge? I assume you will find a solution quite quickly – but if you, do can you explain why this should be so?

Solution on page 104

GREAT MINDS THINK ALIKE!

Plato (c.428–c.348 BC) Greek philosopher, who founded the Academy at Athens

Plato considered the abstract speculations of pure mathematics to be the highest form of thought of which the human mind was capable. He therefore had written over the entrance to the Academy "Let no one ignorant of mathematics enter here."

TIME TO LEAVE

In temperate zones, deciduous trees shed their leaves in the autumn. The trigger for this appears to be the shortening length of the days, even if the weather is still quite mild.

Surely it would be in the trees' interest to keep their leaves until the very last moment – in other words, until the temperature fell to a dangerously low level. Can you find an explanation, considering that Mother Nature is the eternal professional in these matters?

Solution on page 104

HOT AND COLD

The rotation of the Earth has some strange effects on our physical environment. Can you, for instance, explain why during winter it is so much cooler in New York than in Madrid or Naples although all three cities are on nearly the same latitude (about 40°)? More surprisingly still, New York is even colder than London, which is much farther north (51½°).

Solution on page 105

THREE POINTS

The illustration shows an angle and a point P, chosen at random within the angle.

An indefinite number of lines can be drawn through P, cutting the angle in two points. One of these lines will intersect at B and C respectively so that BP = PC, making P the centre of BC.

Prove that this is the line which makes the area of triangle ABC the smallest possible.

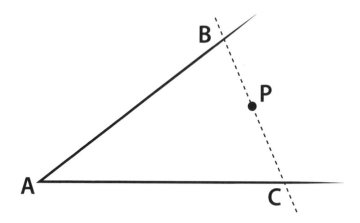

Solution on page 105

STRAIGHT LINES

Suppose we have a number of flat objects, each with a different shape: square, circle, and so on. If we select any of those objects we could draw a bisector across it. We can define a bisector as a straight line that divides an object into two halves. Of course, for any object we could draw an infinite number of bisectors. Would all the bisectors necessarily go through the same point for any specific object?

Solution on page 106

MOVING ON UP

Whilst holidaying in Austria last year, I was pleased to find that many cable cars were still working, even though it was summer. It provided an easy way of getting up and down the mountains. One of the lifts had only two trains, each of five gondolas, continually moving up and down, so that there was always one train ascending and one descending.

I happened to be in the leading gondola, travelling downwards, and watching the other train moving up towards us. As the two trains must be equally balanced, it occurred to me that I ought to be able to tell when the train was exactly halfway down.

Would this happen when my gondola passed the first, the middle, or the last gondola of the other train?

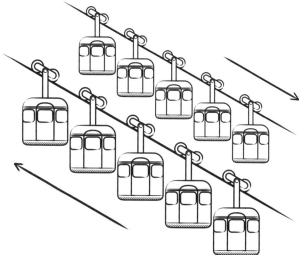

Solution on page 107

MIRROR, MIRROR

It is often said that mirrors reverse left-right but not up-down. Can you think how a single plane mirror can:

1. reverse up-down as well?
2. reverse up-left and down-right?
3. reverse up-down but not left-right?

Solution on page 107

EYEBALL TICKLER

What are the tiny spots floating in your line of vision before your eyes? At first you might try to brush them away until you realize that the specks must be in your eye. "What can it be?" you ask yourself.

Is it a small particle on the surface of your eye? Is it a fragment of something on the retina or am I merely imagining it? If I'm not, is it dangerous?

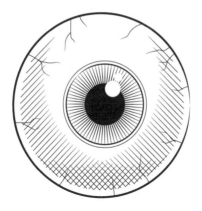

Solution on page 107

SOAP DRAMA

I live in an area where the water is rather "hard". This means that it is quite difficult to get soap to lather and bubble up. I have also noticed that after the soap has been used for some time, it gets progressively more difficult to raise a lather.

I can think of three possible explanations. How about you?

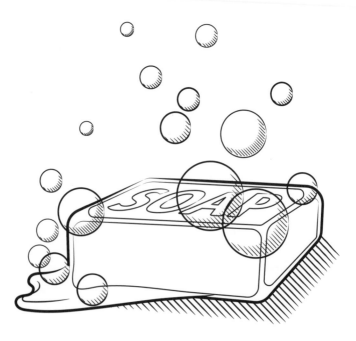

Solution on page 108

LOGO-A-GO-GO

A graphic designer produced a new company logo, as shown.

It consisted of a rectangle with one diagonal drawn. From one of the other two corners two lines were drawn to the midpoint of the opposite two sides. The designer noticed that these last two lines seemed to divide the diagonal into three equal parts.

Prove that the lines drawn from a corner of any rectangle to the midpoint of the opposite two sides trisect the diagonal drawn between the corners adjacent to the original corner.

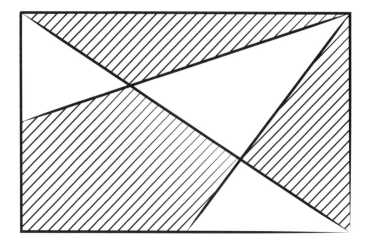

Solution on page 108

WHAT GOES AROUND

Prove by logical reasoning that any three points chosen at random on a sphere must lie on the same hemisphere.

Solution on page 109

GREAT MINDS THINK ALIKE!

Pythagoras (born c.580 BC), Greek philosopher

Seeing a puppy being beaten one day, Pythagoras took pity, saying, "Stop, do not beat it; it is the soul of a friend which I recognized when I heard it crying out."

BLOWING
SMOKE

"Tobacco Seriously Damages Your Health," says every pack, and I agree. The physiological effects on smokers are now well documented and there is little doubt of the detrimental effect of long-term exposure.

However, there are some interesting chemical and physical effects also associated with cigarettes. For instance, when watching smokers, I have noticed that the smoke rising from a cigarette is usually blue, yet when smoke from the same cigarette is exhaled – by blowing a smoke ring, for example – the smoke is always grey. How do the lungs change the smoke's colour?

Solution on page 110

COOL ENERGY

Many of us have to watch our waistlines and take into account the calorific value of the food we eat. In other words, we have to balance the energy contained in our food with the energy we need to live. It would seem a good idea if we could have tasty food that had no calories in it at all. In fact, food technologists have produced an oil which cannot be digested, but no-one can yet be certain what the long term effect of eating it would be.

A friend of mine told me that celery is the only food that requires more energy by the process of digestion than you get back from the food itself. I do not know whether this is true or not, but it set me thinking. An ice cube must have a negative calorific value, because we must give it energy to melt it in our mouth, and this produces water that has no calorific value.

I am quite partial to the occasional packet of peanuts. Why don't I suck a few ice cubes, to nullify the effect of the nuts?

Solution on page 110

HIGH FLYERS

We know that as the radius of a circle increases, so does its circumference. This must mean that the higher an airplane flies, the greater the distance it has to fly to its destination.

On a flight between London and New York, an airliner has to fly an extra three kilometres (two miles) because of its high altitude.

Forgetting air-space congestion as a possible argument, why do airliners not fly by a shorter, lower route?

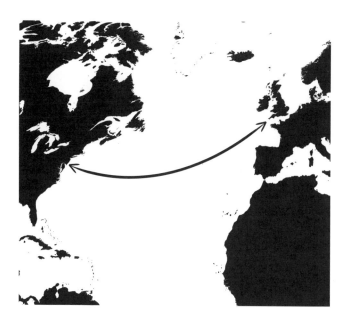

Solution on page 111

TO THE LIMIT

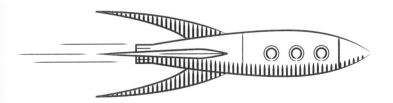

Cars have a maximum speed. If a driver keeps the accelerator pressed down, the speed increases up to that maximum. By contrast, a rocket that fires its engine in outer space will increase its speed, no matter how fast it is already going.

Why does a car have a maximum speed, but a rocket does not?

Solution on page 111

ONCE UPON A TIME

A long time ago two captains of sailing ships had a wager with each other as to which was the faster way to sail around the world: east to west or west to east. They decided to put it to the test by having a race. So, one day at exactly the same time, they set sail from a small island each going in the opposite direction.

Some months later they happened to arrive back at the same island at exactly the same time. They were just about to decide that the bet was off when they compared logs and found that there was a discrepancy of two days between them.

Is it possible to account for the discrepancy and decide who circumnavigated the world faster?

Solution on page 111

ACCELERATION

If a car is parked facing downslope, a marble placed on the floor will roll towards the front. If a car accelerates along a horizontal road, a marble will roll towards the rear of the car.

Would it be possible to accelerate a car down a slope at such a rate that a marble would remain stationary on the floor?

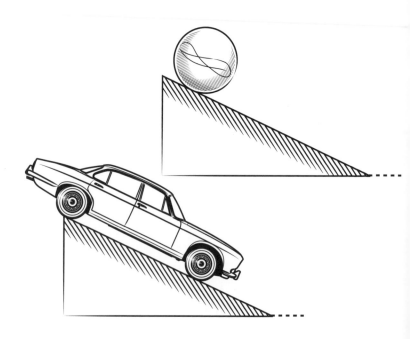

Solution on page 112

GOOD
VIBRATIONS

Seismographs are sensitive instruments for detecting earthquakes. They consist of a large mass suspended by springs. When an earthquake occurs, the disturbance travels through the Earth, making the instrument vibrate. However, the mass tends not to move because of its inertia. This difference in movement is amplified and written out as a trace.

Can you explain why most places in the world receive two traces for each single earthquake event? Can you also explain why a few places detect only one?

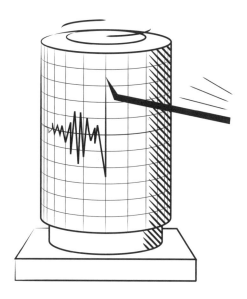

Solution on page 112

FIT TO BURST

There used to be a trick popular in Victorian times. A beer barrel was completely filled with water. A long thin tube had been attached to the top of the barrel, which was empty at the start of the trick. By pouring a very small amount of water into this tube the barrel could be made to burst, demonstrating that small causes can have large dramatic effects.

Suppose one jug of water was needed for the trick. If a narrower tube with half the cross-sectional area had been used, how much water would have been needed to burst the barrel?

Solution on page 113

IN FREEFALL

If a plane stalls, the pilot will be unable to pull out of the dive immediately, but is likely to succeed after having nosedived for some time.

Why?

Solution on page 113

SNOW DRIFT

Have you noticed how much more snow is proportionally deposited on the sides of posts and poles rather than on the sides of buildings?

Why is this?

Solution on page 113

THIRTY-INCH RULER

Hold a ruler horizontally, balanced across your index fingers.

Now, try to move your fingers together so that both slide. It does not work. The ruler first slides over one finger and then over the other, but never both together.

Can you figure out why?

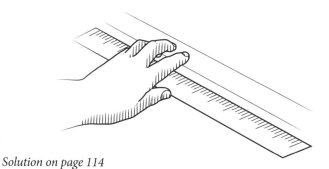

Solution on page 114

GREAT MINDS THINK ALIKE!

Gerolamo Cardano (1501–1576), Italian mathematician and astrologer

Cardano was renowned throughout Europe as an astrologer, even visiting England to cast the horoscope of the young king, Edward VI. A steadfast believer in the accuracy of his so-called science, Cardano constructed a horoscope predicting the hour of his own death. When the day dawned, it found him in good health and safe from harm. Rather than have his prediction falsified, Cardano killed himself.

TARGET PRACTICE

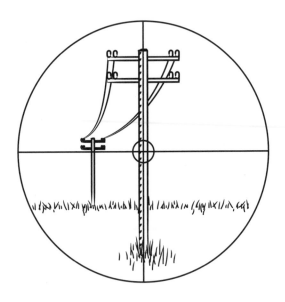

You are standing on top of a high-rise building in Greenwich with a long-range, high-powered rifle, perfectly aligned. You are trying to hit a telegraph pole in Louth, Lincolnshire. Assume that you can see the target through the telescopic sight on your rifle. Would you aim straight at the target, to the left, or to the right of it?

To give you a clue, both Greenwich and Louth are on the 0° meridian.

Solution on page 114

GASPING FOR AIR

A great deal of research has gone into diving and the physiological aspect of ascending to the water's surface. Imagine that you are using scuba equipment at a depth of 25 metres (80 feet), when you find that the equipment is suddenly faulty. You are forced into a rapid ascent, hoping that the air in your lungs is sufficient to take you to the surface.

Would you, in such an emergency, gradually release air as you ascend, thus reducing your air reserves, or would you hold your breath?

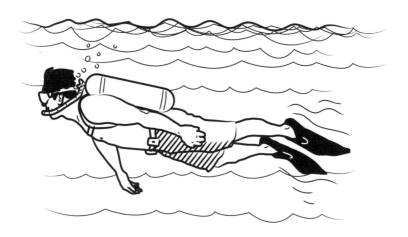

Solution on page 114

BOOMERANG

The boomerang is a wooden missile used mainly by Australian aborigines. It was originally used as a weapon, but tossing the boomerang has recently become a sport. It is generally held vertically in the right hand, although some are designed for left-hand use. The fascinating aspect of this weapon is its ability to return to the hands of the thrower – no matter how far it is thrown.

Can you explain this phenomenon?

Solution on page 115

FIZZY BUBBLY

When the French and English broke through to join hands at the centre of the Channel Tunnel, the workers and local dignitaries celebrated the event by opening several bottles of the finest champagne. To their disappointment the bottles hardly popped and the poured champagne was flat. Yet when the people returned to the surface, they felt sick and could not stop burping.

Why?

Solution on page 115

ROCKET WAR

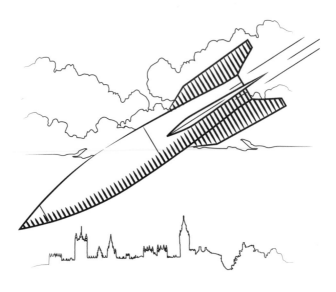

Those of us in Great Britain who lived through the Second World War will remember, without much nostalgia, V-1 and V-2 rockets replacing the bombing raids that had become too expensive for the Germans in terms of losses in men and planes.

It is in our nature to be adaptable. We soon found that the V-1s presented no danger as long as you could hear them. However, as soon as the buzzing of the engine stopped, it was advisable to dive under the bed. No such strategy was available with the V-2.

Why not?

Solution on page 116

THE FLAG

Assume that the wind blows with uniform force from the same direction. A weather vane will remain stationary while a flag, even if perfectly smooth and fully spread by hand to start with, will flap once released.

Why?

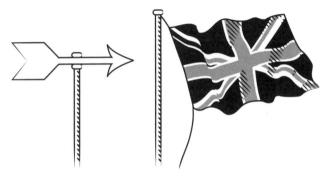

Solution on page 117

Solution on page 117

GREAT MINDS THINK ALIKE!

René Descartes (1596–1650), French philosopher and mathematician
Descartes's coordinate system was one of his main contributions to the development of mathematics. It is said that the idea came to him during a period of idleness in his military service as he lay on his bed watching a fly hovering in the air. He realized that the fly's position at every moment could be described by locating its distance from three intersecting lines (axes). This insight was the basis of Cartesian coordinates.

BOAT ON
A LAKE

Can you steer a boat on a lake if it is entirely calm and you have no oars? The answer must be no.

Imagine the circumstances are the same except that your boat is drifting in a fast-flowing river. Can you then steer with the rudder?

Solution on page 117

HIGH FLYING BIRDS

Do you ever ask yourself what it is that enables birds to fly? After all, they are heavier than air and therefore do not float.

To fly they require lift and forward propulsion. How do birds achieve this?

Solution on page 118

UP IN THE AIR

There is an old legend of an escapee from an island prison who evaded capture by swimming underwater, using a metal pipe he had stolen from the prison workshop as an air tube.

Sharks apart, is the story feasible? And is there a limit to the depth at which the swimmer can breathe through the tube?

Solution on page 118

SCORCHER

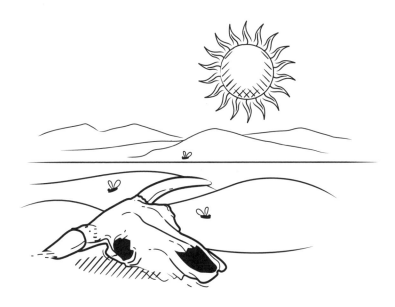

Death Valley is a desert region in southeastern California, USA. Most of the valley is below sea level and it has the distinction of being one of the hottest places in the world. In 1913 a temperature of 56°C (134°F) was recorded – the highest ever, at the time.

Elementary physics has taught us that hot air rises and cold air sinks. Would you therefore not have expected Death Valley to be a cool or moderately warm place, particularly as it is almost entirely enclosed by mountain ranges?

Solution on page 119

HOLE IN
THE BOTTOM

Drill a number of small holes in a can and fill it with water.

1. Fit an airtight lid and the water stops leaking from the holes. Why?
2. Drill a hole in the lid and the water now flows. Why?
3. Remove the lid, run your finger through the leaking streams, and , if the holes are not too far apart, they will merge as shown, forming a single stream even after you have removed your finger.

Why?

Solution on page 119

PANAMA CANAL

The construction of the Panama Canal is regarded as one of the greatest technical achievements of all time. It was completed ahead of schedule and was in full operation by the summer of 1914. With the map of the world in mind, one would have no doubt that the Canal runs from west to east Surprisingly, the Pacific end lies somewhat east of the Atlantic end.

There are some other interesting aspects to which the reader is invited to find an answer:

1. At the last lock, as the gate is opened, any ship will move out to sea without tugs and without using its own power. What makes it move?
2. One would assume that the water levels in the Atlantic and Pacific are the same. However, there is a difference at times of as much as 30 centimetres (12 inches). Why are the ocean levels not the same?

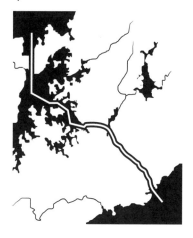

Solution on page 119

THE WORLD'S BEST SCIENTIST?

Let's give your brain a quick break to relax and recover from all the head-scratching and pay tribute to the greatest scientist who ever walked the planet, the genius who ripped up the rule book – Sir Isaac Newton (1642–1727), the renowned English physicist and mathematician.

Sir Isaac Newton invited a friend to dinner but then forgot the engagement. When the friend arrived, he found the scientist deep in meditation, so he sat down quietly and waited. In due course dinner was brought up – for one. Newton continued to be abstracted. The friend drew up a chair and, without disturbing his host, consumed the dinner. After he had finished, Newton came out of his reverie, looked with some bewilderment at the empty dishes, and said, "If it weren't for the proof before my eyes, I could have sworn that I have not yet dined."

THE HOURGLASS

You have an hourglass on the table in front of you, which has run its course. What happens if you turn it upside down? Will the weight of the hourglass be reduced while some of the sand is in free fall?

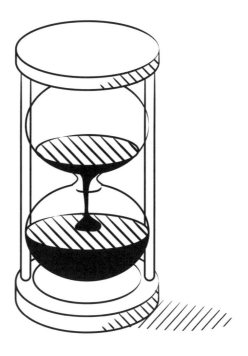

Solution on page 120

TOY BALLOONS

You have two identical balloons. Balloon A has been blown up to 10 centimetres (4 inches) in diameter and Balloon B to 20 centimetres (10 inches).

You want to blow 10 times into each balloon. Will it be easier to succeed with one balloon or the other, or will the necessary force be the same?

Solution on page 120

TELESCOPIC VISION

We are very used to the idea of telescopes and binoculars, but what are they doing? Do they make distant objects look closer or bigger?

What about a magnifying glass: does it make things look bigger, or is it doing something else?

Solution on page 121

THE UNBELIEVABLE TRUTH

I was reading an advertisement the other day. It had been placed in a newspaper by a firm of energy consultants, and showed a diagram of an average house to illustrate the following figures: 50 per cent heat loss through the walls, 30 per cent through the doors and windows, and 20 per cent through the roof and floors.

The firm claimed to be able to cut each of these percentages by half. What is wrong with this claim?

Solution on page 121

STARING AT THE SUN

Ordinary sunglasses work by absorbing some of the light that would otherwise be transmitted. Polarizing sunglasses work in a slightly different way: they will transmit only light that is vibrating vertically. As light reflected from wet roads, water, and so on tends to vibrate horizontally, this cuts down glare from such surfaces. If you are wearing a pair of polarizing sunglasses and hold up another polarized pair rotated through 90 degrees, the lenses appear black because no light can now pass through both sets of glasses.

Could you insert a third set of polarizing sunglasses between the first two in such a way as to permit some light to be transmitted?

Solution on page 122

A CUP OF TEA

There is nothing more cheerful than the sound of a kettle coming to the boil and the prospect of a refreshing drink to come. Have you noticed that the kettle goes strangely quiet just before the boil begins, and can you explain why?

Solution on page 122

HOT WATER

I could never understand why, whether at home or in a hotel, after turning on the hot water the flow of the stream first decreases, only to return to normal after about 30 seconds. Do you know what causes this?

Solution on page 122

SPEED RACER

You have driven a car for many years, and fancy yourself an expert. You are driving along a highway at 110 kilometres per hour (70 miles per hour). The weather is pleasant and you have not a care in the world. Suddenly you see a small animal trying to make it across the road. You cannot avoid it by swerving because there are cars on your left and right. You are lucky that there are no cars closely behind you because you are an animal lover and the only way to avoid killing the poor thing is to brake.

Will you slam on the brakes and lock them, or will you apply increasing pressure hoping you can stop just in time?

Solution on page 123

HOLE IN ONE

In the early days of golf, the balls were smooth. Dimples were introduced later, after manufacturers claimed the dimpled variety travelled farther. Were they right, and if so why?

Solution on page 123

Solution on page 123

GREAT MINDS THINK ALIKE!

Leonhard Euler (1707–1783), Swiss mathematician

When Euler first came to Berlin from Russia, Frederick the Great's mother, the dowager queen Sophia Dorothea, took a liking to him and tried to draw him out on a number of topics. Euler, no courtier, replied in monosyllables. "Why," asked the dowager queen, "do you not wish to speak to me?" Euler replied, "Madame, I come from a country where, if you speak, you are hanged."

COLOURLESS

Clean air, water, and transparent glass are supposed to be colourless. Yet the sky is blue, a sheet of glass looked through sideways is green, and a mountain lake, pure as it is, appears blue.

Textbooks in physics will talk about Rayleigh scattering, electric field, and electron oscillations, but there is a simpler, if unscientific, explanation. Do you know what it is?

Solution on page 123

SOAP BUBBLES

One of the most beautiful sights in our physical world is that of soap bubbles showing all colours of the rainbow. Is the air pressure inside the bubble larger or smaller than, or equal to, the atmospheric pressure outside the bubble?

Logic will provide the answer.

Solution on page 124

Solution on page 124

GREAT MINDS THINK ALIKE!

Carl Friedrich Gauß (1777–1855), German mathematician

Someone hurrying to tell Gauß that his wife was dying found the great mathematician deep in an abstruse problem. The messenger blurted out the sad news. "Tell her to wait a minute until I've finished," replied Gauß absently.

SPRAY GUN

When we were children, it was fun to annoy grown-ups by spraying them with water.

We blew through a horizontal tube to force water up a conduit tube into a fine spray. The same principle is used in aerosols and paint sprayers. Why does water rise in the tube, against gravity?

Solution on page 124

BICYCLE PUMP

Here is an easy one, to give your neurons a rest. When you pump up a bicycle tyre by hand, the valve gets hot and so does the hand-pump on the downward stroke. If, however, you use a compressed-air bottle instead, the bottle cools.

Can you explain the different effects?

Solution on page 125

GLUE STICK

What makes glue stick?

You will probably say that this is easy to understand, but difficult to explain. If pressed further, you might ascribe it to some chemical property of the adhesive. This is just begging the question, and furthermore is wrong. The answer is rather more complex.

Have another guess.

Solution on page 125

QUICKSAND

Many years ago I saw a Western in which a cowboy and his horse sank into this mass of loose wet sand and were unable to extricate themselves. The bulging eyes of the man and his desperate attempts left an indelible impression on me.

I wondered what the physical properties of quicksand were, what caused the bulging eyes and whether there was a method which would offer the best chance of survival.

Solution on page 125

PROTECTIVE SHIELD

A German epic poem, the *Nibelungenlied*, written about AD 1200, tells the story of Siegfried, a prince from the lower Rhine, who is determined to woo Kriemhild, a beautiful Burgundian princess.

Siegfried has become famous for killing a dangerous dragon, whose blood has turned his skin into a shield impenetrable by any weapon. However, unnoticed by him, a leaf from a tree covered a tiny spot on his shoulder blade, leaving him vulnerable.

The question is, could a personal defensive shield become a scientific reality?

Solution on page 126

AEROPHOBIA

Many people hate everything about flying. Taking off, landing, the noise of lowering the undercarriage and the change of pitch on acceleration or deceleration of flying speed.

While much of it is standard procedure, I am at a loss when we fly through a thunderstorm with lightning all around us. In fact, I am a little concerned myself about that.

Why is there no catastrophic damage if lightning strikes a plane?

Solution on page 126

CENTRIFUGAL FORCE

I watched an Olympic hammer-thrower on TV recently, and the presenter was explaining how the heavy weight was thrown outwards by "centrifugal force – the same force that tries to throw you outwards as a car turns a tight corner". This brought back memories of my schooldays, and I could almost hear my old physics teacher saying, "There is no such thing as centrifugal force."

This set me puzzling. Perhaps my teacher was wrong, or just behind the times.

Does centrifugal force exist? If so, what is the source of the force?

If not, why do so many people talk about it as if it did exist?

Solution on page 127

LINES ON THE GROUND

I was recently on a summer walking holiday in Austria. Being one of the less enthusiastic ramblers, I often took the ski lift to the top of a mountain, and then walked downhill. Whilst sailing over the alpine meadows one day, I noticed a number of horizontal ridges round the mountain. They were about a foot apart, and a few inches in height. For a few moments, I thought that they were perhaps contour lines. I asked a geographer friend of mine what they might be, and received a rather complex explanation involving something called "soil creepage" and gravity.

Can you think of a more prosaic reason?

Solution on page 127

FEEL THE BURN

The amount of calories you shed is near enough equivalent to the energy you use up in performing a physical task. Suppose you walk briskly for eight kilometres (five miles), and on another occasion you run the same distance.

(a) Would you expend the same amount of energy?
(b) Assuming that the walk takes two hours, would you use up the same amount of energy if you ran for two hours?

Solution on page 127

SCALES

The other day, I was watching two small children playing. One had a toy crane, and the other had some pretend kitchen scales. This set me wondering. Suppose the crane was fixed in one of the pans in such a way that the end of its arm was above the other pan, and the pans then rebalanced. If I placed a weight in the non-crane pan the balance would, of course, move that way; but what would happen if the weight were suspended from the crane? Would the balance behave as if the weight had been put in the pan below its point of suspension, or as if it had been placed in the pan with the crane?

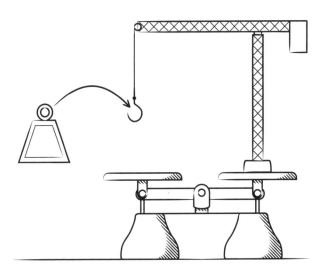

Solution on page 128

SHORT OF WORK

In October 1997, researchers at The Royal Free Hospital London announced the result of research conducted on 6,500 people born in 1958. They had found a positive correlation between lack of height at aged seven and the chances of being made redundant subsequently in life.

This does seem to be a baffling result; why does being short at age seven increase the chance of being made redundant years later? Can you think of a mechanism to cause this? Should we be giving growth-promoting hormones to children to cure the economic ills of the country?

Solution on page 128

SHOCK TACTICS

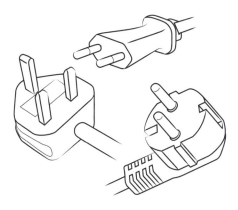

There are no electrical standards that the whole world is prepared to adopt. This means that when travelling between countries, you have to be prepared to take all sorts of adapters and to cope with different voltages.

In the United States, for instance, the voltage is 110v, whilst in most of Europe, it is 220v. Does this mean that European electricity is twice as powerful than American? Is American voltage safer?

Solution on page 129

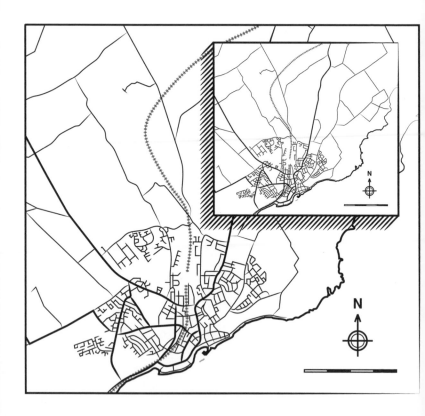

Solution on page 129

THE PUZZLE

Suppose I had two maps of exactly the same region, but of different scales, so that one was much smaller than the other. Imagine that the smaller one is sitting on top of the larger one at a random position, but in such a way that it is entirely within the confines of the larger one, and with its axes aligned parallel to the larger map.

What are the chances that there is a point on the smaller map that lies exactly over the same point on the larger map?

If you can answer that question, you might like to consider whether the chances would be different if:

1) the smaller map was upside down
2) the axes were not aligned
3) the smaller map was crumpled up

Assume each time that there is no part of the smaller map that overhangs the larger map.

GREAT MINDS THINK ALIKE!

(Jules) Henri Poincaré (1854–1912), French mathematician and philosopher of science

Poincaré once gave a lecture at the Sorbonne. A female student, wanting to congratulate him, reported that the hall had been so crowded that she had been unable to get a seat. "I had to stand throughout your lecture." "So did I, Mademoiselle," replied Poincaré.

DIG THIS

We have all seen how carefully archaeologists have to be when they are digging out ancient remains and artefacts. We all know that the further down they dig, the further back in time they are going. Some discoveries can be buried very deeply indeed.

By what processes do these old ruins manage to bury themselves in the ground? Why do they not stay on the surface?

Solution on page 130

DRIFTING BY

Clouds are so familiar, that we hardly give them a second thought. But we are all familiar with the fact that there are many different sorts, and sizes. Just for a moment, think of a summer's skyscape with a few clouds drifting by. If you have a similar image to me, you will have imagined all the clouds with their bases at about the same height above the ground. In fact, the height of the cloud base is of great importance to the pilots of small planes.

Why, and how, do clouds align themselves so that their bases are roughly level?

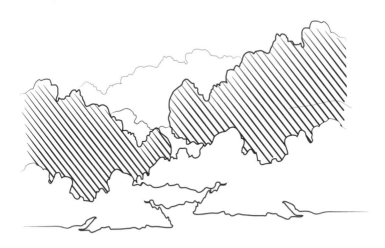

Solution on page 130

BULLET TIME

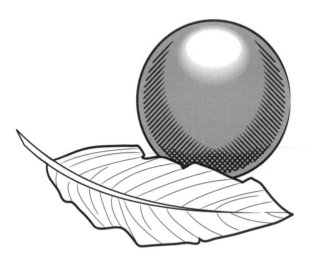

It is axiomatic that the acceleration in free fall due to gravity is the same for all bodies. Although this is not borne out by experiments in the atmosphere, we know that this is due to air resistance. Drop a bullet and a piece of paper simultaneously, and the bullet will hit the floor fist.

If you take the thought further, you will agree that if you drop the same object, perhaps a steel ball, from a great height, first during winter with the air temperature at 25°F, and then at a temperature of 80°F, in the latter case the ball will hit the ground in lesser time owing to lower air density and consequently less air resistance.

Now suppose you have a swimming pool and you repeat the experiment, first with the water temperature at 25°F, and then at 50°F. At which temperature would the object hit the bottom of the pool in a shorter time?

Solution on page 130

VIRTUAL PARTICLES

The November 1997 issue of *Mensa Magazine* contains a fascinating article on subatomic particles written by Colin Wagstaff.

It proposes an analogy that presents an interesting problem.

Before putting this to the reader, it might be worthwhile to recap the article.

The Greek scholars thought of the atom as the smallest particle in nature. This proved too simplistic a theory. Further work by nuclear physicists discovered a different picture, namely:

(a) An atom consists of a nucleus with orbiting electrons.
(b) The nucleus itself was made up of protons and neutrons.
(c) Protons and neutrons were in turn made up of quarks.
(d) Quantum electrodynamics (QED) deals not only with intermolecular theories but has also postulated that the universe is not a sterile empty space but "…is constantly spawning new particles which appear suddenly and then disappear, existing for a time so unimaginably short that they were christened 'virtual' particles."

To explain their effect, Wagstaff suggests the analogy that I have turned into a challenging puzzle.

Picture two boys each in his own boat in a lake: The boats are about one metre (three feet) apart and one of the boys throws a heavy rucksack to the boy in the other boat.

(continued on page 82)

Questions:

1. What happens to the water level of the lake while the rucksack is in the air?

2. What happens to the two boats?

Solution on page 131

NORTH POLE

Porto Alegre in Brazil is situated on meridian 50° west and latitude 30° south, at a distance of approximately 15,000 kilometres from the North Pole. Eucla in western Australia is also 15,000 kilometres from the North Pole.

What are the odds that Eucla, or for that matter any other spot on Earth equally distant from the North Pole, is more than 15,000 kilometres from Porto Alegre?

Solution on page 131

Solution on page 131

GREAT MINDS THINK ALIKE!

David Hilbert (1862–1943), German mathematician

The mathematician George Polya likes to tell stories about Hilbert's absentmindedness. At a party at the great mathematician's house, his wife noticed that her husband had neglected to put on a clean shirt. She ordered him to do so. He went upstairs; 10 minutes passed; Hilbert did not return. Mrs Hilbert went up to the bedroom to find Hilbert lying peacefully in bed. As Polya puts it, "You see, it was the natural sequence of things. He took off his coat, then his tie, then his shirt, and so on, and went to sleep."

PARACHUTES

In the early days of test jumping, it was found that parachutes had a tendency to swing wildly from one side to another. This occurred even in calm weather, creating in some cases a condition that could lead to the collapse of the parachute.

Why should this be the case, and what simple, but ingenious idea, completely eliminated the danger?

Solution on page 131

FORMULA ONE

I was never particularly interested in motor racing. One of my friends, an enthusiast, hoping to convert me, explained the basic principles.

However, in spite of my friend's powers of persuasion, I remained sceptical because I believed that the major factors in success were luck and the mechanical superiority of one car against the others, with the expertise of the driver having only a minor effect on the outcome.

My friend then explained that the driver's skill was decisive. If, for instance, the trailing car overtakes the leading car in an expert manner, he can create a push forward for his car and at the same time slow down his adversary.

Why should this be so?

Solution on page 132

CLAY PIGEONS

I recently took up Clay Pigeon Shooting. This involves trying to hit skimming clay disks with a shot from a shotgun. There are an amazing range of circumstances; the clay may be left or right, fast or slow, high or low, climbing or descending, at right angles, approaching or retreating. However, whatever the situation, I always seemed to miss.

I did notice that whenever I had a crossing clay (passing at right angles), my instructor told me that I missed by shooting behind if it was high, and in front if it was low.

Can you think why this might be so?

Solution on page 132

TOP SKATERS

Ice skaters often finish their presentations with a dramatically fast spin on the spot. Presumable they do this at the end so that any resulting giddiness will not affect anything else they might try to do. The skater can speed up the spin by bringing in their outstretched arms.

I am interested in this from an energy point of view. The extra energy involved in the faster spin presumably comes from the muscular effort involved in drawing in the arms. But if the skater stretches out the arms again, the spin slows once more.

Where does the energy go when the skater slows down the spin?

Solution on page 133

FLYING BACKWARDS

I am one of those people who always try to get a window seat when flying, and spend most of the time gazing out. There is always something interesting to see.

The other week I had a seat just near the leading edge of the wing. Looking down I could see a car travelling in the same direction along a road that ran parallel to the direction we were flying in. We were, of course, travelling much faster than the car and so the wing soon overtook it and the car fell behind underneath the wing. I correctly worked out that if the car had been travelling faster than the plane, it would have drawn away, and if the two speeds had been the same we would have maintained our relative positions.

Sometime later the plane was turning, and I looked along the edge of the wing and was surprised to see the tip moving backwards relative to a stationary object on the ground. Did this mean that the air was flowing the wrong way over the end of the wing, and if so how did it keep flying?

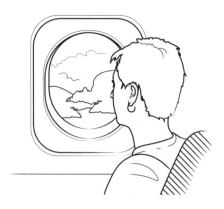

Solution on page 133

DOWN THE PLUGHOLE

A friend of mine recently visited an African country through which the Equator runs. He told me of a tourist attraction where the direction of rotation of water flowing down a plughole was demonstrated to be different in the North and South hemispheres. An entrepreneur had set up three basins, one a few yards north of the equator, one south and one actually on the Equator. He then proceeded to show that in one the water rotated clockwise as the basin emptied, another moved anticlockwise, and the one on the Equator did not rotate at all.

My friend thought that this was a convincing demonstration of the Coriolis effect, and tipped the guide handsomely, like most of the other tourists.

I am more sceptical. Can you think why?

Solution on page 133

BISECTING
A TRIANGLE

ABC is a general triangle and P is a random point on the hypotenuse. Q is a midpoint between AC. BP connects P with the opposite vertex. QR is a line parallel to BP. Now connect P with R and prove that triangle ARP bisects triangle ABC.

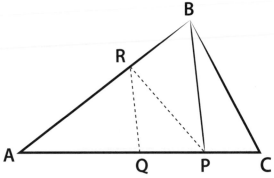

Solution on page 134

Solution on page 134

SHUTTLE
SERVICE

Alan and Bert were neighbours. Every day they would walk to the train station together, and there they would wait on different platforms, as they were going in opposite directions to work. Alan always envied his neighbour because Bert had only a few stops to go away from town and always got a seat, whereas Alan had a long trip into town and the train had already filled up at the nearby terminus.

Alan noticed something else that puzzled him. It seemed that Bert's train arrived first more often than his. Alan knew that the train service consisted of a couple of trains that plied back and forth and there was no timetable. And they arrived at the station at random times.

Alan decided he must be imagining it, but he started to keep records. To his surprise he found that Bert's train was 50 per cent more likely to turn up first. How did Bert manage this?

Solution on page 135

THE GLOBE

I have an antique globe in my study, which is unique inasmuch as the surface area and the volume are both four-digit integers in inches, times π.

Find the radius.

Solution on page 136

ANGLES IN A SEGMENT

It is said that all angles in the same segment are equal. ABD and ACD are two triangles in the same segment. Can you prove that the angles at B and C are equal?

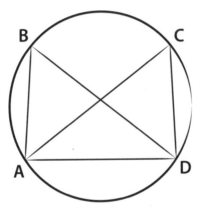

Solution on page 136

GREAT MINDS THINK ALIKE!

John von Neumann (1903–57), American-Hungarian scholar

John von Neumann was an incurable practical joker. During the Second World War, when he constructed his electronic brain for the government, he identified it on delivery as a Mathematical Analyser, Numerical Integrator, and Computer. Scientists worked with it for several days before they realized that the first letters of the name its inventor had given it spelled MANIAC.

MONUMENT SQUARE

You are walking along a road, looking at a statue on top of a pedestal. As you approach it, you notice the statue getting larger to the eye until you reach a point after which the size diminishes.

You know from the tourist guide that the statue is ten feet high, standing on an eight-foot plinth. Prove that the statue will appear tallest when you have reached a point twelve feet from the bottom of the plinth.

For the sake of clarity we accept point E as being the eye of the observer and point C as being the centreline of the statue at road level.

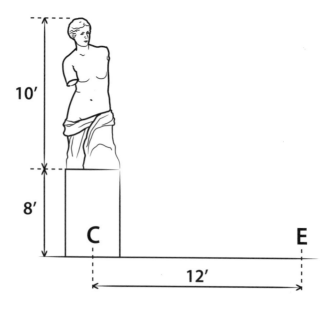

Solution on page 137

THE SCIENTIST

Square ABCD has sides each of 20 inches. Lines are drawn from midpoints E, F, G and H to opposite corners to form square HIJK. Prove that this centre square has an area of 80 square inches.

Use inspiration rather than arduous calculations.

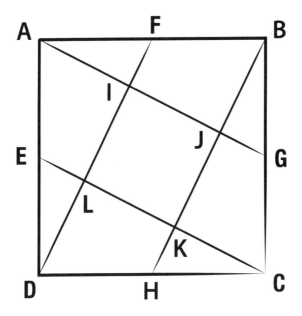

Solution on page 138

IN THE SHADE

Archimedes postulated a formula for the shaded area as: $A = \dfrac{\pi x^2}{8}$

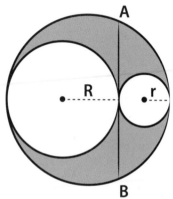

where x is the length of line AB.

Prove that the formula is correct!

Solution on page 139

TRACK AND FIELD

Take a circular track bounded by two concentric circles. Using the Zero Option, prove that the area of the track is equal to the area of the circle whose diameter is a chord of the larger circle and a tangent to the smaller circle.

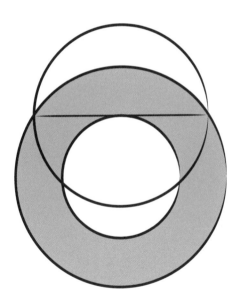

Solution on page 140

ANSWERS

How Do You Survive On An Iceberg?

Polar ice does indeed contain salt. Such ice, melted down, is as undrinkable as seawater. However, over time, the brine in the ice blocks will migrate downward, because of gravity. This draining effect will make the melted ice drinkable after about a year, and it will be almost completely free of salt after several years.

The problem does not affect all Polar Regions, as some ice is formed by precipitation. However, many areas are because there is insignificant snowfall, and such little as there is will he blown away by winds of up to 160 kilometres per hour (100 miles per hour).

Among the various desalination techniques, freezing has been developed as an alternative method, based on the different freezing points of fresh- and seawater, but the equipment needed is beyond the reach of Eskimo communities. If you had the equipment when you were stuck on the iceberg, you might just make it!

Head Scratcher

In 200 BC, Eratosthenes devised an ingenious method to measure the distance around the Earth. He was aware that at noon during the summer solstice in the city of Syene (in Egypt) a vertical rod did not cast a shadow, while in Alexandria (5,000 stadia = 500 miles away) the vertical rod cast a shadow which formed an angle of 7°12'. With this information he was able to calculate the circumference of the Earth to within 2 per cent of its actual value.

A School of Fish

A swimming fish creates vortices as shown in the figure below:

The vortex directly behind the fish causes the water to flow in the opposite directions of the fish's motion, while to the left, right, above and below the vortices flow in the same direction as the fish. The school will instinctively position itself so that the energy needed is minimized.

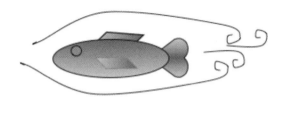

Tree Paradox

The circulation of sap was first investigated in a series of experiments by Stephen Hales at the beginning of the eighteen century. It is a great mistake to think of trees as a hollow tube through which water is either drawn from the top by the process of transpiration, or pumped up the plant by the roots. The tree is a living organism and there are subtle processes at work inside the living cells which move water up, and sugars produced by photosynthesis, down the plant.

For evidence that cells can perform such amazing feats, look no further than ourselves. In the process of osmosis, water passes through semipermeable membranes to dilute a more concentrated solution. Yet inside our kidneys, cells ensure that the opposite happens, recycling water to prevent dehydration.

Knuckles

The finger joints are lubricated by a fluid in which tiny gas bubbles are embedded. By pulling the fingers, you change the pressure in the joints and upset the equilibrium, the bubbles collapsing/re-dissolving, resulting in the characteristic crack.

Some time is needed before the gas bubbles are available in the fluid once again, for a repeat performance.

Under Pressure

Blood pressure, apart from other factors, varies also with the distance from the heart. In order to standardize readings, they are taken at heart level. Otherwise the height of the patient would become a factor and the readings would become unreliable.

Seeds of Truth

Coconuts are, of course, seeds of a type of palm tree. The radicle, which would develop into a new plant is located under one of the three "eyes" at the end of the coconut. Before transportation, a hot iron is placed against each of the eyes to kill the radicle and prevent the coconut from germinating. This lengthens the shelf-life of the nut.

Pest Control

Researchers have found that in situations where there are hunters (ladybirds) and hunted (greenfly), the population tends to vary in a cyclical manner. If unchecked there is an explosion in the number of the hunted (point A, on the graph), which provides plenty of food for the hunters, whose numbers also increase (B). The hunted population then collapses (C) because of the predation, causing an eventual fall in the number of predators also (D). And so on.

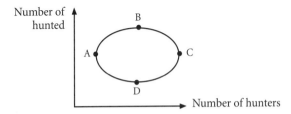

In fact, depending on the starting position, there will be a series of such "population loops":

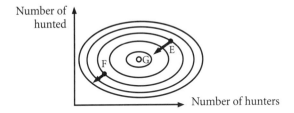

If you happened to spray when both populations were high (E), then the population moves to a smaller loop and the numbers remain low. If you were unlucky enough to apply the insecticide when the populations were low, you could move the situation to an outer loop, meaning that you would eventually have more greenfly, than if you had done nothing.

However, we said that the situation was stable, so we must have been in the centre of the pattern (G), and therefore spraying would be bound to make the situation worse.

Boxing Match

Aerodynamics will always make the match fall with an edge leading, as this offers the least air resistance. On hitting the floor, the match is bound to turn on its flat side because of the location of the centre of gravity. If, however, you bend the match to form an angle, then the centre of gravity will be located within the angle formed by the bend.

Time To Leave

Thousands of years of evolution have ensured that trees shed their leaves at the optimum time. If they drop their leaves too early, then valuable time would be lost. If the leaves fall too late, they could be damaged by frost, and this would be detrimental to the plant as a whole.

Hot and Cold

Warm sea-water rises above cold water and, because of the higher centrifugal force at the Equator, the upper, warmer, current to the North Pole deviates to the right towards Europe, while the colder counter-current passes along North America.

Three Points

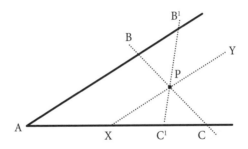

We know by premise that triangle ABC is the smallest in area. Now let us prove it. First draw an arbitrary line B^1C^1 by rotating BC through P clockwise. Now apply the Zero Option principle by reducing the angle AB^1C^1 to zero. You can do this by moving B^1 to infinity, turning B^1C^1 into line XY parallel to AB. This must clearly be the maximum triangle with infinite area. Turning XY counter-clockwise the triangle becomes progressively smaller up to a critical point, after which it increases again until YX is parallel to AC, once more forming a triangle of infinite area. Somewhere between the two extremes will obviously be the smallest triangle. We also see that, as we turn XY counter-clockwise through position B^1C^1, the difference between B^1P and C^1P diminishes until it reaches zero before it turns negative. Logically, the smallest triangle will be

formed when P is the midpoint. As complex as the explanation might appear, it is "elegant" because it depends on a logical progression, rather than on complicated mathematical operations such as calculus.

Here is a more traditional approach:

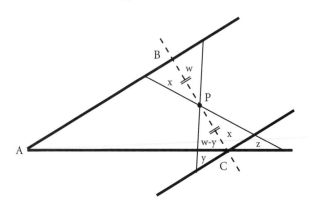

Construct a line through C parallel to AB.

a) Imagine BC rotating clockwise about P. The resulting triangle will increase by W but decrease by $(W - y)$, i.e. its area increases.

b) Imagine BC rotating anticlockwise about P. The resulting triangle will increase by $(X + z)$ but decrease by X, i.e. its area also increases.

Therefore the original triangle's area must have been at a minimum.

Straight Lines

For symmetrical objects the bisectors will be coincident, but this is not true for asymmetrical objects.

Moving On Up

Each gondola would be halfway down as it passed the corresponding gondola of the other train. As I was in the first gondola, this would occur when it passed the other leading gondola. However, the trains would be halfway down when the middle gondolas passed and that would be when my car passed the last car in the other train.

Mirror, Mirror

The answers are:

1. Place the mirror on the floor (or ceiling).
2. Place the mirror at a 45° angle to the floor.
3. I cannot think of one either.

You just have to think of environments where there are large numbers of grazing animals to realize how successful this strategy is for grass.

Eyeball Tickler

Floaters are shadows cast on the retina by microscopic structures in the vitreous humour, a jelly-like substance behind the lens.

Although everybody gets floaters at some time, they are not serious. However, if you have persistent or worrying visual disturbances, you should have it checked out by an expert.

Soap Drama

a) Soap naturally dries out, so the longer it lies around in the air, the drier it gets and the more difficult it is to produce a lather.

b) As the soap gets smaller, the ratio of the surface area to the volume increases. This will increase the drying-out process.

c) As the soap is used, it gets smaller and its surface area decreases. The ease of lather production will depend on the size of the surface area.

Logo-A-Go-Go

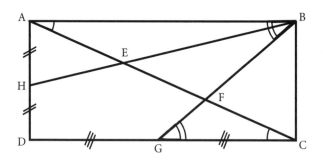

Consider triangles ABF and CGF:

> Angle FAB = Angle FCG (alternate angles), and
>
> Angle FBA = Angle FGC (alternate angles)

Therefore triangles ABF and CGF are similar.

Therefore:

$$\frac{AF}{CF} = \frac{AB}{CG} = \frac{2}{1}$$

and

$$AF = 2CF$$

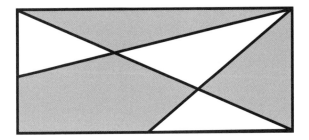

By the same argument:

$$EC = 2AE$$

What Goes Around

The key to the proof is the axiom that any three points must lie on a plane. Visualize a plane through the three random points and you will realize that they must be on the same hemisphere.

Blowing Smoke

The smoke rising from the cigarette consists of extremely small particles, whose size is of the same order of magnitude as the smallest wavelength of visible light. It is the blue end of the spectrum that has the shortest wavelength, and so this light is scattered whereas the longer wavelength light (red light) tends to be unaffected. The smoke particles in exhaled air have had their size increased because of a layer of moisture that has condensed on them. Therefore all wavelengths get scattered equally, giving the smoke its light grey colour.This phenomenon is known as the Tyndall effect, and explains why the sky, which contains many small dust particles, is blue and the sunsets are red.

Cool Energy

Although theoretically sound, this scheme is impractical. The amount of energy needed to melt a gram of ice at freezing point and to warm it up to body heat is about 0.4 Joule (0.1 calorie). The amount of energy obtained from a gram of peanuts is about 25 Joules (6 calories). To nullify a 150 g packet of peanuts would therefore require me to suck 9,0000 cubic centimetres of ice.

High Flyers

Even taking the extra distance into account, it is still cheaper to fly at a higher altitude where the aircraft is flying above the turbulent weather patterns. Also, the thinner air presents less drag, meaning less fuel is consumed. Finally, it may be possible to take advantage of the fast winds found at that higher altitude.

❧❦❧

To the Limit

The car stops accelerating when the thrust of its engine equals the drag of resistance or friction. A rocket in outer space experiences no friction, so there is no limit to its acceleration.

There is, however, a maximum speed at which anything can travel; so the rocket will never exceed the speed of light.

❧❦❧

Once Upon a Time

If they both set off at the same instant and also returned at the same time, then both their journeys would have taken the same time. However, the two ships would have counted different numbers of passing days. If the island had experienced the passage of X days, the ship travelling west against the rotation of the Earth would have counted X − 1 days because it would have made one less rotation than the world. The ship travelling with the spin of the globe would have counted X + 1 days.

This anomaly was corrected by the adoption of the International Date Line.

❧❦❧

Acceleration

Assuming that the marble is frictionless, it will always roll down the slope with the same acceleration relative to the slope. If the car is accelerating at a rate less than the marble's, the marble will roll towards the front. If the car's acceleration is greater than the marble's, the marble will roll towards the back of the car. When the two accelerations are exactly the same, the marble will be stationary relative to the car.

Good Vibrations

There are two ways in which earthquake waves can travel around the world. First, they can travel across the Earth's surface, rather like waves travel across the sea. Second, they can travel directly through the centre of the Earth. These waves travel at different speeds and have different distances to travel, so they arrive at the seismograph at different times.

The Earth does not have a uniform density. The denser core acts like a lens, focusing the vibrations at some points while leaving other places in a "shadow".

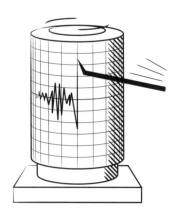

Fit to Burst

The trick works because the water pressure at one point depends on depth, not the weight of water above. If the tube connected to the top of the barrel is narrow, a small amount of water can cause a large increase of pressure in the barrel by increasing its depth below the surface.

If a narrower tube were used, then even less water would be needed to increase the depth. If the tube were half the area of cross section, then half the amount of water would be needed to attain the same height in the tube. So half the jug would be sufficient.

In Freefall

A plane can stall when it is travelling too slowly for its wings to provide sufficient lift. When that happens, it starts to fall. If, as with most planes, there is more weight at the front, it will fall nose first. As it picks up speed, air flows over the wings, once more providing lift. Eventually there will be enough lift to enable the pilot to pull out of the dive.

This also accounts for the characteristic looping flight that you can see with a paper airplane.

Snow Drift

The wind driving the snow diverges many metres in front of a large building, thus dispersing the snow before it hits the wind-side face. A smaller object does not divert the wind, permitting the snow to build up.

Thirty-Inch Ruler

The finger that slides first is the one with less friction between it and the ruler. The friction depends on the weight of the ruler on the finger. Try as you might, you can never balance it precisely. A minute difference is sufficient to favour one finger against the other. However, as the sliding finger approaches the middle, the weight on it increases; therefore, it will stop and the other finger will begin to slide.

Target Practice

You should aim to the left of the target. Any moving object will be deflected to the right north of the equator, and to the left south of the equator, in relation to the rotation of the Earth.

This phenomenon is called the Coriolis effect, named after the French physicist Gaspard de Coriolis (1792–1843), who first analyzed it mathematically. The Coriolis effect is of great importance to meteorologists, navigators and the military.

Gasping For Air

As one ascends to the surface of the water, the external pressure on the body decreases. The lungs contain air that expands and could cause rupturing. Divers are instructed that, during an emergency rise to the surface, they should leave their mouth open, allowing the expanding air to escape.

Boomerang

The return boomerang has a length of 30–75 centimetres (12–30 inches), curving to the left, and capable of more than 90-metre (295-foot) throws. There have been several attempts at explanations. T.L. Mitchell in 1846 suggested that it was caused by the skew combined with the spinning motion. This does no more than beg the question. A more convincing explanation is offered by E Hess in "The Aerodynamics of Boomerangs" (*Scientific American*, Nov 1968).

The boomerang is an airfoil and therefore subject to a lift, which is greater on the top half because it is turning in the same direction as the boomerang, whereas the bottom half is turning in the opposite direction.

Fizzy Bubbly

At the bottom of the shafts the atmospheric pressure was much greater, so that more of the carbon dioxide remained in solution to be released at the lower pressure on the surface.

Rocket War

The V-2 rockets exceeded the speed of sound and therefore you could not hear their approach. You could hear the detonation only after they hit the target, too late to take evasive measures.

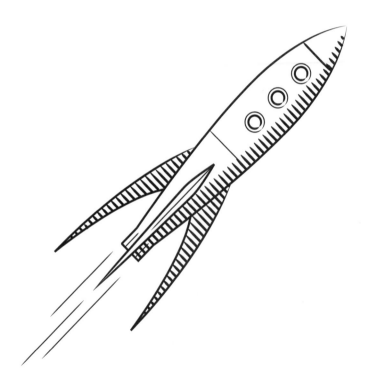

The Flag

The Bernoulli Effect takes over. No flag is perfectly smooth. The tiniest imperfection will make the airflow speed up as it crosses the ripple, reducing air pressure on the ripple side, increasing it on the other side, making the flag flap.

Boat On A Lake

The rudder will have an effect only if there is relative motion between the boat and the water. This answer begs the question "Is there such relative motion?" Most likely yes. However, there are so many forces acting upon the boat that this question cannot be answered with absolute certainty.

First, there is gravity. Imagine the river to be frozen. Ignoring friction, the boat would slide down the slope. Then there is buoyancy, resulting in a component force downriver, partly counteracted by drag. Air resistance adds to the drag, while a wind upstream or downstream has an effect. Furthermore, the river flows at different rates in the middle and near the banks.

It is theoretically possible that all these forces combine to synchronize the motions of the boat and river for a limited period, which would make steering impossible.

High Flying Birds

Lift is easily explained; it is the same principle as applied to aircraft, involving the Bernoulli Effect.

Look at a typical wing, be it that of a plane or a bird: the air above the wing has to cover a greater distance and therefore travels faster, reducing the pressure. There is greater pressure below the wing; both effects produce lift. This enables the bird to glide and soar through the air.

It is more difficult to explain what provides the forward propulsion. High-speed photography shows that a bird's wing changes shape, flexing in a complex fashion, providing a backward thrust to the air during part of its cycle. As a reaction to this the bird is pushed forward, in a way analogous to a swimmer using the breast stroke.

Certainly a straightforward flapping would be ineffective. Neither do I accept Storer's explanation that the twirling of birds' feathers acts as a propeller.

Up in the Air

The story is perfectly feasible, provided the swimmer keeps no more than 1 metre (3.28 feet) below the surface. Any deeper and the water pressure would prevent chest expansion to enable the lungs to inhale.

Scorcher

A number of factors contribute to these extreme conditions. The mountain range on the west side rising to more than 3,300 metres (10,830 feet) absorbs the moisture of the west winds, which, on descending east of the Rockies, are adiabatically heated and dried, turning the valley into a hot, and desert.

Hole in the Bottom

And the three answers are:

1. The pressure in the can is less than the pressure outside. The inside pressure is caused by a few centimetres of water while the outside pressure is caused by the atmosphere, which is equivalent to about 10 meters of water. Water will not flow from a low- to a high-pressure area, as that would be like flowing uphill.

2. By drilling a hole in the top of the lid, the atmospheric pressure is applied to the surface of the water; the pressure outside is still atmospheric, but the pressure inside is now atmospheric plus a few centimetres of water. The water therefore flows out.

3. Surface tension prevents the water streams from separating again. Surface tension is a phenomenon by which the molecules at the surface of a liquid are held tightly by the cohesive forces of the molecules beneath. Water has quite a high surface tension.

Panama Canal

The two responses desired are:

1. The canal is fed by a number of freshwater lakes, including Gatlin and Miraflores. When the last gate leading to the ocean is opened, the freshwater level will still be higher than the denser saltwater, creating a

flow to equalize levels and providing a drift for the vessel.

2. The salinity of the Pacific is higher than that of the Atlantic, and therefore denser, which accounts for the lower level of the Pacific.

The Hourglass

The weight of the hourglass will not change, although some of the grains of sand will be weightless, being in free fall. This will be balanced by the impact force when the grains hit the bottom. There may be some slight fluctuating at the beginning and end of the process, however.

Toy Balloons

It is a surprising scientific fact that as a balloon gets larger its internal pressure drops. It will therefore be more difficult to expand the smaller balloon. We have all experienced how hard it is to start inflating a balloon, after which it becomes easier. In fact, the point at which it starts to get easier (and the internal pressure begins to decrease) is when the balloon has inflated to approximately 1.4 times its uinflated diameter. A very simple experiment can demonstrate the fact. Try gently blowing up two identical balloons simultaneously, one partially pre-inflated. You will find that the larger one will inflate in preference to the smaller one.

Telescopic Vision

If a telescope (or a pair of binoculars) is being used in the correct way, it is said to be in "normal adjustment". When utilized in this way, the eye is perfectly relaxed when using it. When a perfect eye is relaxed, it focuses at infinity. A telescope does not, therefore, give you an image that is closer. It does, however, make the object look bigger.

A magnifying glass increases the focusing power of the eye, allowing the object to be placed closer. There are many circumstances in which it works by not magnifying at all. Try this simple experiment to see what I mean. Hold a piece of writing so close to the eye that it cannot be seen clearly because it is blurred. Now insert a magnifying glass between the eye and the writing. You should now find that the writing is clear enough to read but is no larger than it was.

The Unbelievable Truth

Notice that the total heat loss is 100%. This is always so, however well insulated the house is; all the heat put in will eventually get out until the temperatures inside and outside are equal. What is important is the rate at which heat escapes. Doubling the insulation of the house will halve the rate of heat loss, and thus also halve *the rate* at which heat will have to be supplied to maintain the same temperature. This would also halve the energy bills.

But the proportion of heat loss through the walls, windows, and roof would remain roughly the same.

Staring At the Sun

Strangely enough, the answer to this question is "Yes". If there are just two crossed polarizing filters, no light will pass through the combination. If you insert an extra filter between the other two, at 45° to them, then some light will travel through this triple combination. In fact, the amount of light transmitted through the triple combination is about half of what would pass through a single filter.

This is because some light will always manage to get through a double combination at 45° to each other. So some light will pass through the first and second filter. Some light will also pass through the second and third filter. Therefore, some light will pass through the complete triple combination.

A Cup of Tea

Where the water is being heated, it is locally hotter than elsewhere. Small bubbles of water vapour will form, which collapse as they move to cooler parts. The "singing" sound is caused by these thousands of tiny collapsing bubbles. When the water is close to boiling, the bubbles survive, causing the kettle to go quiet.

Hot Water

The answer is so simple that I should have known the first time I noticed. The hot water heats the tap valve first, which expands, reducing the flow. After the whole tap has warmed up, the effect disappears.

Speed Racer

Do not lock the brakes, for there is less friction between the tyres and the road if they are sliding rather than rolling. Under normal road conditions you would require about 25 per cent less distance to stop if you kept the tyres rolling.

It is very difficult to judge exactly how much pressure to apply to the brakes to maximize deceleration while avoiding locking. Many modem cars have anti-lock braking systems that detect when the wheels are about to lock and automatically momentarily release the brakes.

Hole in One

In fact, dimpled golf balls travel about four times farther. The effect is caused by the backspin imparted to the golf ball by the dub. As the top of the ball spins backwards, it drags air which would otherwise have travelled beneath the ball. This air has to speed up to travel the extra distance, and this causes lift in a very similar way to an aircraft wing. If the ball were smooth, the spin would have no effect.

Colourless

There is no substance that is perfectly transparent. Light is gradually absorbed as it passes through an increasing thickness of any substance. That is why it is completely dark at the bottom of the ocean. If all colours are absorbed equally, there is no resultant colour. Ordinary glass has a slight tendency to absorb magenta, leaving the green hue seen through thick glass.

The sky appears blue because that colour has the shortest wave length and it is scattered by small dust particles in the air. There can be two reasons why mountain lakes look blue. First, the water is reflecting light from the sky. Second, water from melted glaciers is often coloured because of the high mineral content.

Soap Bubbles

The air pressure inside the bubble is greater than the atmospheric pressure outside. It could not be less or it would collapse.

The surface tension necessary to form the bubble is directed towards the centre, trying to reduce the size, and works therefore in tandem with the atmosphere. The inside pressure must, therefore, be greater to balance the two forces acting against it.

Spray Gun

The air travelling through the blow tube at considerable speed reduces the air pressure in the conduit tube. The liquid in the container is subject to atmospheric pressure and is therefore forced up, and out of, the conduit tube in the form of a fine spray.

Bicycle Pump

The resultant temperature changes are caused by a phenomenon known as adiabatic process. In addition to an explanation, a brief definition of what is termed an adiabatic process is in order. Such a process is one in which no heat (at least in the short term) is entering or leaving a system. If gas is compressed in a cylinder – for instance, a bicycle pump – it undergoes an adiabatic *change* and heats up. If gas is released, as in the compressed air bottle, the expansion results in the cooling of the gas.

Glue Stick

Molecules of a substance attract each other by a process called cohesion. Molecules will also attract molecules of a different substance, and this process is called adhesion. The trick is to find substances whose molecules are very adhesive. To work effectively the molecules have to be in close proximity, which is why adhesives are usually liquid and the surfaces involved should be as clean as possible. Rather surprisingly, water is quite a good adhesive. If, for example, you wetted two pieces of wood, placed them together and put them in a freezer, the pieces of wood would be very difficult to part once frozen. Theoretically, if one could hone two surfaces to a degree that molecular contact was possible, no adhesive would be needed. In practice such surfaces are contaminated by dust and other impurities, so that adhesives are needed.

Quicksand

Quicksand is a physical state in which saturated sand loses its supporting capacity when mixed with water and assumes the character of a liquid.

Under these circumstances a person may become engulfed as in a fluid.

Theoretically the body should float, as quicksand's density exceeds that of humans, but struggling will shear the quicksand and make it more difficult to free oneself. The best tactic is lying down on your back and rolling towards terra firma. The eyes bulge because of the hydrostatic pressure on the lower part of the body and the sheer terror of the situation.

Protective Shield

Not as an equivalent to the dragon's blood. However physics can have a similar effect, by stopping an offensive weapon, though it could hardly be called a protective shield. This is one way it may work: if, say, a metal sword is swung back and forth into a magnetic field, electric currents are generated in the metal, which causes it to lose energy by heating.The heat is dissipated, diminishing the kinetic energy until the sword eventually stops. It is doubtful whether the effect could be made to work fast enough to protect the intended victim.

Aerophobia

Michael Faraday discovered that there is never any electrical charge inside a hollow conductor – this is called the Faraday Cage Effect. This is why car radios often fade when driving underneath power cables, or steel bridges. In fact, planes get struck by lightning quite frequently, and the charge passes harmlessly round the surface of the aircraft. Many passengers would probably not even be aware of the incident. Pilots do try to avoid thunderclouds, however, not because of the lightning, but because of the ferocious up- and down-draughts that can occur inside them. This could cause structural damage to the plane and interfere with the flow of air over the wings. One is often given the advice not to shelter under trees during an electrical storm. Sitting in a car would be much safer.

Centrifugal Force

The teacher was correct: there is no such thing as centrifugal force. Moving objects will move in a straight line, unless acted on by an unbalanced force. To make an object move in a circle, there must be force acting *towards* the centre, i.e. a centripetal force. For the Moon, this is supplied by the Earth's gravity, and for the athlete's hammer, this is provided by the thrower pulling inwards. If the wire should break, the hammer would not fly outwards, but would continue in the direction it had been travelling in at that instance.

Lines on the Ground

The ridges are caused by cattle grazing. They are likely to align themselves horizontally, as walking round the mountain is easier than up or down. The first time this will make small grooves if the ground is soft. On subsequent occasions, the cattle are likely to use these ridges again, as they will afford greater purchase. Thus the ridges are reinforced.

Feel the Burn

(a) Yes, provided you walk reasonably briskly.
(b) No. You will use much more energy if you run instead of walk for the same period of time. This can be compared to the consumption of petrol for a car travelling five miles and twenty miles respectively.

Scales

Think in terms of the combined centre of gravity (c.o.g.) of the beam, pans and their contents; that point where all the weight could be considered to be concentrated. If the c.o.g is vertically above or below the pivot for the scales, then it is balanced. If the c.o.g. is to the right, the pans move clockwise, and if to the left, they move anticlockwise. Adding the weight as shown would move the c.o.g. to the left, so the balance would react as if the weight had been put in the pan below its point of suspension.

Short of Work

There is an easy trap that many people can fall into (and, to be fair, the researchers in this case did not). If event A happens before event B, and they are related, it is obvious that B cannot have caused A, so it seems reasonable that A must have caused B. This is not necessarily correct – there could be a third event C, which might have caused both A and B. This is what occurred here. The researchers discovered that if people are subjected to long-term stress early in childhood, this stunts growth, and can cause instability in adulthood, which could well lead to difficulty in holding down a job. In a similar way, other people have noticed a correlation between truancy and later criminal behaviour, and are therefore taking action to improve school attendance in the confident expectation that this will have a beneficial effect on the future crime rate. We shall have to see whether this proves to have the expected result.

Shock Tactics

Firstly, we need to point out that any device connected to a voltage that it is not designed for, is almost certain to be seriously damaged. Electrical power is determined by the combination of current and voltage. This means that in the USA, devices must use twice the current to produce the same power as the corresponding European device. Therefore, the wires have to be thicker, otherwise they would heat up unacceptably. American voltage is safer in the sense that if you were to receive an electrical shock, it would drive only half the current through you. Unfortunately, this can still be enough to be fatal.

The Puzzle

The answer to the first question is that there will always be a point on both maps that will be coincident. There are various mathematical proofs of this, but we can solve this quite easily with a 'Thought Experiment'. Maps are representations of an object drawn to some scale. We did not say what the scales of the maps were, and this must be irrelevant. So let us suppose that one of the scales is 1:1. In other words, we could work with the small map and the country that it represents. Let us say it is France. Now I can rephrase the original question like this:

"If you take a map of France to some random point in France, what are the chances that the point you are standing at is represented somewhere on the map?" Put like this, the answer must be 100 per cent. If I also asked if your answer would be different if the map had been upside down, or rotated, or crumpled, I am sure that you would still answer, "No".

Dig This

There are probably a number of factors at work. Firstly, there is some settlement of heavy objects in soft soil, as natural ground water moves soil particles from underneath. Secondly, there is continual wind erosion in some parts, and so wind will continually settle dust over everything, and dirt will be washed out by rain. Thirdly, vegetation will shed leaves and other material that will gradually degrade into a type of soil, and plants will grow over low obstacles and, as years go by, over each other, and build up a thickness of soil.

Drifting By

It is a mistake to consider clouds to be solid objects, like pieces of cotton wool. Clouds consist of billions of droplets of water that have condensed out of the air. The condensation occurs when the temperature of the air drops to a point where the humidity is 100% or greater. This point will depend on the original humidity of the air, and the temperature gradient above the ground. At a certain height, the conditions will be just right, and clouds will form.

Bullet Time

At 50°F. At 25°F the steel ball would not hit the bottom at all because the water would be frozen.

Virtual Particles

1. There must be a vertical component to the rucksack's trajectory. Therefore the throwing boy would apply a force downwards on his boat, making it dip into the water a little more, causing the level to rise. The boat would then bob up and down for a little while (probably longer than the flight of the sack) about a mean level lower than previously. On catching the rucksack, the other boat would also bob up and down, with the lake eventually returning to the original level.
2. The two boats will drift apart.

North Pole

The odds are zero. In other words, the distance cannot be larger. For Eucla, or any other spot, to be 15,000 kilometres from the Pole it must be on the same parallel as Porto Allegre. But the circumference of the Earth at the Equator is near enough 40,000 kilometres and only slightly less (because of the flattening effect) around the Poles. Therefore the maximum distances between any two locations on the 30° parallel, via the South Pole, is approximately 10,000 kilometres.

Parachutes

On descending, air fills the parachute, producing a high-pressure area inside. Air then starts to flow round the 'chute, creating unstable vortices, encouraging air to tip out of the canopy at unpredictable point round the circular edge. Creating a small central hole allows a stable airflow, enabling the parachute to descend steadily without reducing its overall drag very much. You can demonstrate a similar effect for yourself. If you cup your hand and then force it through water in the direction of your palm, you can feel your hand being moved at sideways. Opening the fingers makes the movement much smoother.

Formula One

To start with, the trailing car experiences less of a drag because of the reduced air resistance in the leading car's flow shadow. When the trailing car begins his overtaking manoeuvre at the optimal distance from the side of the leader, the air is forced through the narrow space between them, reducing the pressure (Bernouilli effect). The trailing car has then greater pressure from behind, causing acceleration, while the leading car will experience a corresponding deceleration.

Clay Pigeons

The secret of successful Clay Pigeon Shooting is to aim in front of the clay. This is because the shot takes time to travel, so you have to find a point in front of the clay where clay and the shot will arrive together. The faster the clay is moving, the further in front you have to aim. So it all depends on the sportsman's accurate perception of speed. When the clay is high, there are no nearby objects, and most novices tend to underestimate its speed. This causes a "lead", which is too small, and the shot passes harmlessly behind the clay. Low clays are often seen against a background of trees, etc., and this tends to cause an overestimation of its speed, and hence a shot passing in front.

Top Skaters

There is no change in energy. The rate at which anything spins is determined by the distribution of mass relative to the axis of rotation. When the arms are drawn in, more mass is near the axis and as a result the skater automatically spins faster. Imagine a weight on a string tied to the top of a pole. As the weight rotates round the pole, the string gets shorter and the rate of rotation round the pole automatically increases.

Flying Backwards

The plane and its wings will at all times be moving forward through the air. Let us imagine that the plane is flying in an anticlockwise circle. It is easier to imagine what is going on if we consider this to be equivalent to the plane being stationary, and the ground rotating under it in a clockwise direction. There will be a centre of rotation (c.o.r.), and anything on the ground between the c.o.r. and the plane will appear to be moving back relative to the plane. Anything beyond the c.o.r. will appear to be moving forwards.

Down the Plughole

The Coriolis effect is very difficult to demonstrate in practice. There is a very small tendency for the rotation of the Earth to effect the way in which water rotates as it drains, but this is easily swamped by other effects, like the disturbance caused by pulling out the plug. So how was my friend duped? Probably by employing another property of water, in that it has a tendency to "remember" disturbances. For instance, suppose you swirl water vigorously in a given direction, and then leave it to settle. The water will tend to drain in that same direction when released some time later – even long after the water appears perfectly calm again. Try it for yourself.

Bisecting a Triangle

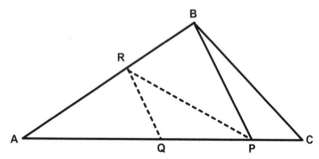

Given AQ = QC and RQ is parallel to BP

BPR = BPQ (same base, equal height)
BXR = XPQ (BPX is common) ... (1)

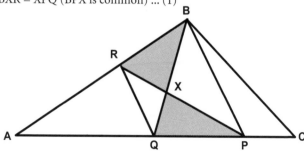

ABQ = QBC (equal base, same height) ... (2)
From (1) & (2)

ABQ – BXR + XPQ = QBC – XPQ + BXR   or
ARP = PRBC
i.e. RP bisects ABC

Shuttle Service

Bert's advantage is that he was travelling towards the nearby terminus. It is slightly easier to picture if you think of an equivalent situation. Suppose the travelling trains were represented by arrows moving round a circle, see below. Alan and Bert's positions are represented by two dots, and both the terminus and the town would be equidistant from the two men. Lines joining their positions to the centre would divide the circle into two sectors. If both trains were in the larger sector, Bert's train would arrive first.

If one train were in the smaller sector, then Alan's train would arrive first.

It can be seen by inspection that there is a higher chance that Bert will get his train first. If you want to do the maths, I think you will find that if the distance to the town is four times the distance to the terminus, then the chances of Bert's train arriving first is 3:2.

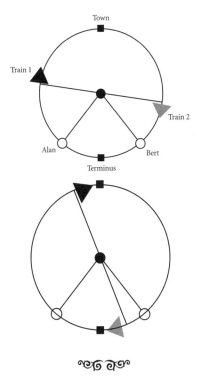

The Globe

To meet the "four-digit" condition, $4r^2$ and $\frac{4}{3}r$ must lie between 1,000 and 9,999. As far as $4r^2$ is concerned, r must be larger than 15 and smaller than 50.

In respect of $\frac{4}{3}r^3$, r must be larger than 15 and smaller than 20.

If r is to be an integer, it must be divisible by 3. Only 18 meets this condition, and therefore $r = 18$.

Angles in a Segment

If you can prove that the angle AOC at the centre of the circle is twice the angle ABC of any triangle in the same segment, then you have also proved that all these angles are equal.

Triangles AOB and COB are both isosceles (radius of the circle). Let angle ABO be x and angle CBO be y. Then angle AOD must be 2x and angle COD must be 2y. In other words the angle subtended at the centre by AD is always twice the angle at B subtended by AC.

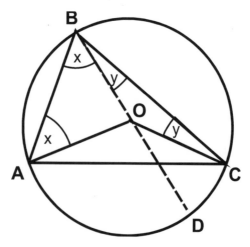

Monument Square

The statue will appear tallest if angle AEB is widest. Draw an imaginary circle through A and B, being tangent to the road RR at E.

The angle subtended by AB is by definition the same on all points of the circle. This angle outside the circle is smaller. It is larger inside the circle but is beyond the eye of the observer. Only point E meets both conditions.

As DB = 5 ft, ME = MB = 13 ft, DM = CE = 12 ft (by Pythagoras).

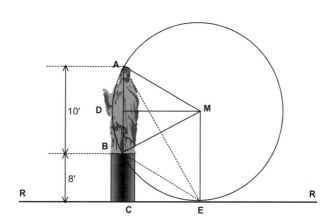

The Scientist

Constructing the shaded triangles, we have added the same area as we have taken away to form the cross AIOBJPCKMDLN, which has therefore the same area as the original square, equal to 400 square inches.

However we now have five small squares, making the centre square 80 square inches.

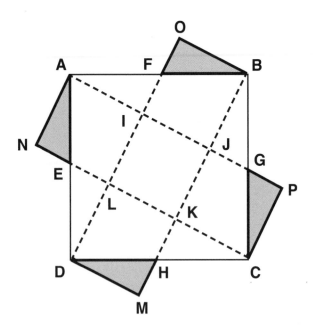

In The Shade

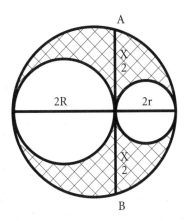

The area of the large circle $= \varpi (R + r)2$

The area of the shaded part is equal to the area of the large circle less the areas of the two smaller circles. Hence:

$$\text{Shaded area} = \varpi(R+ r)2 - \varpi R2 - \varpi r2$$
$$= \varpi(R2 + 2Rr + r2 - R2 - r2)$$
$$= 2\varpi Rr \ \dots(1)$$

Also let the chord AB be x.

Then

$$\frac{\dfrac{x}{2}}{R} = \frac{2r^2}{\dfrac{x}{2}}$$

Therefore

$$Rr = \frac{x^2}{16} \ \dots(2)$$

Combining (1) and (2) gives:

$$\text{Shaded area} = \frac{\pi x^2}{8} \quad A = \frac{\pi x^2}{8}$$

Track and Field

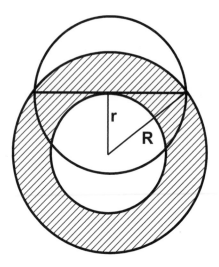

Let the radius of the track's outer circle be R and of the inner circle r. Then the area of the track is $R^2\varpi - r^2\varpi$ or $\varpi(R^2 - r^2)$.

Also the chord's length is $2\sqrt{R^2 - r^2}$ which is the diameter of the circle, whose area is $\varpi(R^2 - r^2)$ equal to the track's area, which we set out to prove. Using the zero option principle, by making the radius of the inner circle zero, the solution is found immediately.

The interesting feature of this problem is the fact that by applying the Zero Option, making $r = 0$, the track and the circle concur.

PUZZLE NOTES

PUZZLE NOTES